PAIDEIA MONOGRAPHS

CHRIST AND GOVERNMENT

S.A. DE GRAAF

www.paideiapress.ca

Christ and Government

This English edition, translated by P.J. Boodt, is a publication of Paideia Press (3248 Twenty First St., Jordan Station, Ontario, Canada L0R 1S0). Copyright ©2023 by Paideia Press. All rights reserved.

Except for brief quotations in critical publications or reviews, no part of this book may be reproduced in any manner without prior written permission from Paideia Press at the address above.

Paideia Monograph Series Editor: Steven R. Martins

Book Design by: Juan Esteban Clavijo

ISBN 978-0-88815-337-1

Printed in the United States of America

CHRIST AND GOVERNMENT

OF LATE IT HAS BECOME quite clear that all problems have not been solved among us yet; on the contrary, there are many questions left. This is the case not only with regard to the church, but also in the political realm. Everywhere we are faced with many difficulties, and discussions about these difficulties are continuously going on.

To state this does not mean that we regret this situation. On the contrary, it is a source of joy for us that there is an active confrontation with the principal questions on all levels. We will never be finished with those questions. But that is not really a serious drawback. We are often faced however with having to make a decision which has a principial background, and then the difficulty arises that this principial background is not sufficiently clear to us. Happily, as believers we are often led intuitively to the right decision, but it will be clear that we may not make our intuition the decisive factor. We would run the danger of showing lack of direction and of following a haphazard policy. For that reason it is necessary, highly necessary, for us to reflect more and more on our principles, and to come to a conscious realization of them.

One of the questions in discussion among us, and one which cries out for an answer is the following: Is the government only subject to God Almighty, the Triune God, the Creator of heaven and earth, or should it also know itself to be subject to Christ, as the King who has been crowned by God?

To this question we have often answered that the government is indeed subject to the sovereign God and therefore also to the eternal Son, as second person of the Trinity, but not to the Mediator, to the Word-become-flesh, the Christ who is also man, who has suffered and died, who was resurrected and now sits at the right hand of God. According to this view the dominion of Christ only extends to his Church. From God as the Sovereign one draws, on the one hand, a line to Christ, to whom the church is subject, and on the other hand, one draws a line to the government, to which the citizens of the state are subject. What Christ is for the Church, the government is for its subjects.

This thought has found acceptance among many of us, and is taught on all hands. This question is the one which we shall discuss in the present essay. I hope to prove from Scripture first of all that we have to confess that the government is subject to Jesus Christ. In the second place, I hope to answer the question of how we have to understand the dominion of Jesus Christ over the state. And finally, I want to say something about the practical consequences of this thesis.

I will begin with the Scriptural proof. Though we are quite conscious of the fact that Scripture does not make political pronouncements, since it is directional and has the intent to evoke faith, we need not think or believe that Scripture would leave us in doubt about a question which pertains to the Lordship of Christ.

I could immediately refer you to the word of Christ himself: "To me has been given all power in heaven and on earth," an utterance which in its sweeping generality tells us very much. However, this pronouncement needs further analysis, which I want to postpone until I have come to the point where I intend to develop the thought about the nature of the dominion of Jesus Christ.

For that reason I want to consider first 1 Timothy 2:1-7. And also here it is necessary to have a closer look. In this passage the command is given that supplications, prayers, and intercessions be made for all men, and especially for kings, and all those who are in authority. This exhortation is insisted upon and made the more pressing by the reference to the will of God that all, i.e., all kinds of men, not only the common man, but also those in authority, shall be saved. This is followed by the further explanation that there is one God who has created all those men, not only the common man but also the man in authority, and that there is one Mediator between God and men, the man Christ Jesus who thus has come for all those people, therefore also for those in authority, and outside of Whom nobody can be saved.

The decisive point here is the significance of this expression; the Salvation of kings and of those in authority. Does Paul simply mean that those in authority will be saved as far as their personal life is concerned, or does he mean that they will also be saved as persons-in-authority, in the exercise of their office, and should learn to subject themselves to God as such? Does Paul only have in mind the salvation of those magistrates, or does he have reference to a future in which magistrates will be Christian magistrates and will govern in a

Christian way?

We have to pay attention to the circumstances of the life of Paul and the congregations at that time. Especially in the Roman Empire the glory of might was worshipped, and was not seen as given by God, but rather was adored as something which itself was divine, and that might, that power, was exercised by the several authorities in an anti-God spirit (anti-Christian spirit). In circumstances like that Paul had to foresee the eruption of conflict between the government and the church. This conflict already became evident in the persecutions.

When he now exhorts the Christians to pray for all those who are in authority, he is saying at the same time that this is according to the will of God, who is also intending the salvation of the government. What has to be prayed for by us, therefore, is not only a certain wisdom for those in authority, but especially their conversion. And the result of that will be that the believers may lead a quiet and peaceful life in all godliness and honesty. Paul's intent, therefore, is the conversion of the magistrates, their turning to God, their breaking with the spirit which until now governed the execution of their task, and would inevitably lead to conflict with the church. The magistrates had to subject themselves to God and learn to serve Him in their governing, so that by their government the Christian life would be furthered.

Paul adds the thought that there is not only the one God, to whom the magistrates are subject, but also the one Mediator, through whom the magistrates must turn to God. In this we thus also find the demand that the magistrates also recognize Jesus Christ as the Mediator through whom they go to God. Without the recognition of Christ, it is impossible for the magistrates to confess that the source of their authority is God. There is no relation between God and the magistrates in the exercise of their authority except through Christ. Because they receive their authority from God through Jesus Christ it is possible for them to serve God in their wielding of authority only through Jesus Christ.

In addition to this, there is Psalm 2 in its entirety in which we are told about the rebellion of the nations and the kings against the Lord and his Anointed. The first thing we have to observe is that the Psalmist speaks of YAHWEH, the God of the covenant, who has covenanted with us in Jesus Christ. Further on we are told about the decree which is made, according to which Christ is anointed as King, and then the kings are called to serve YAHWEH and to kiss the Son, that is, the theocratic king. Here I cite Calvin: "However much therefore, the princes of this world please themselves in their shrewdness, we may know that they are without any wisdom until they have become humble disciples of Christ." Perhaps Professor Noordtzij states it even more clearly in the *Korte Verklaring*: "The Messiah-King of Israel is the Son of the Lord; God himself has set him in his theocratic office. Therefore his is the dominion of the world; He may do with the nations as he chooses." For that matter, when the kings are told here to "Kiss the Son" or as Noordtzij translates, "render homage to the

Son," what other meaning could be intended than that the princes should offer him their fealty, be conscious that they govern only by His will, and owe obedience to Him?

Next I would remind you of the names given to the Lord Jesus Christ in the book of Revelation. If one would not accept as evidence that the expression "King of Kings and Lord of Lords" signifies that he is truly King and truly Lord (cf. Greydanus Comm. On Reveations 17:14) then one certainly has to accept as proof the title, "The prince of the kings of the earth," in Rev. 1:5 (cf. 17, 14:19, 16). The princes of the earth also stand under Him, have to serve Him, to obey Him, and to submit to Him. And this is said not about the Eternal Son, but about "Jesus Christ, who is the faithful witness, the first begotten of the dead."

I hope I have presented sufficient evidence from Scripture for the thesis that the kings and magistrates are subject to Jesus Christ and have to obey Him, or in other words, that Jesus Christ is King also of the magistrates.

Another question is whether the confessions state the same. And we have to answer this question in the affirmative. In the Catechism we confess that Christ sits at the right hand of God the Father, that He might there appear as the head of His Church. Through Him, the Father governs all things, and therefore also the kingdoms and governments. We could hardly interpret this expression to mean that Christ governs those kings in spite of themselves, nor that His dominion does not include the calling for the governments to serve Him and to recognize Him as having the dominion.

In the second place I refer to the prayer after the worship service as found in our hymnals, in which we ask God that the King of kings might reign over them and their subjects. In this prayer the desire that the government might subject itself willingly to the dominion of Jesus Christ is clearly expressed.

As proof to the contrary Romans 13 is often used, where it is said that the government is the servant of God; in this context Christ is not mentioned at all. If one would deduce from this that Christ is not King of the state, then this deduction can only be the result of the thought that it is possible to speak about God on any terrain without speaking about Jesus Christ. And then one has separated God from the revelation in Jesus Christ. In all of Scripture God is no other than the God who reveals Himself to us in Jesus Christ.

It is perhaps needless to say that the dominion of Christ is a derived dominion, that is to say, Christ owes His authority to the Triune God. Thus it is rather evident that when one speaks about the source of authority, as is done in Romans 13, he then speaks about God, and not about Jesus Christ. This is not meant as a denial of the fact that this divine authority over us is exercised by Jesus Christ. I have written somewhere that the magistrates govern by the grace of Jesus Christ. If I had meant by it that the source of the authority lies in the Mediator, I would have erred. I meant, however, that the magistrates govern by the grace of God, which does not come to us except through Jesus Christ.

In the second place I would discuss with you the question of how we have to see this dominion of Christ over the state and the government.

It seems that two objections arise here: first the fact that state and government have been instituted only temporarily because of sin, and second, the related fact that the government has been entrusted with the sword. Those objections seem to gain the more force when we confess the dominion of Christ to be eternal, and to be a dominion of peace. What kind of relationship exists between the eternal dominion of Christ and the temporary authority of the government? And what relationship exists between Christ's reign of peace and the power of the sword which the government has? These are the two questions which need answering.

Let us first speak about the temporary character of the authority of the government in relation to the dominion of Christ. In Art. 36 of the Belgic Confession the statement is made: "We believe that our gracious God, because of the depravity of mankind, has appointed kings, princes, and magistrates; willing that the world should be governed by certain laws and policies; to the end that the dissoluteness of men might be restrained, and all things carried on among them with good order and decency." The confession is plainly that government is there because of sin. We could ask the question whether something of the power which the government now has would have been exercised also if there had been no sin; but in this context we need not answer that question. In its present form the government exists because of sin. What now is the relation between this interim-power of the government and the dominion of Christ? Is its interim character not an obstacle in the attempt to reduce it to the dominion of Christ?

We would perhaps hesitate here if in connection with the dominion of Christ we had to limit ourselves to thinking of that dominion as being only eternal. But that is not the case. Besides eternal dominion He has also been given temporary dominion which He, at the end of days, will put down again in the hands of the Father (1 Cor. 15).

What is the difference between the eternal and temporary power of Christ, and what is the distinctive character of His temporary power? This difference becomes apparent when we compare the power of Christ with the power which Adam had, and which he would have exercised if sin had not come into the world. Without sin Adam would have remained our Head, and he would have served continually in an official capacity which meant not only that all were incorporated in him, but also that he exercised a certain power over all. He would have reigned as head of humanity, although that also meant that all with him would reign over the works of God's hands. This power and authority of the first Adam has been laid in the hands of the second Adam. Thus Christ is an eternal King, and His dominion is an eternal dominion.

As a result of sin however it became necessary that besides that power he should be endowed with a different power, namely a temporary one. Now he had to receive also the power to combat his enemies and to ensure the victory of His Kingdom in this world. This kind of power could never have been given to Adam; it could be given to Christ because he is not only man, but also God. He is the incarnate Word.

And it is with this temporary power of Christ that the power of the government corresponds. Like the temporary power of Christ, the power of the government serves to combat sin. And in that dominion of Christ the government has to serve Him. In developing my third point I will have to indicate the limitations of the authority of the government, but at this point it is my intention only to indicate that in the temporary character of the power of the state no argument can be found to obviate its reduction to the power of Christ.

The same holds for the second objection which was raised. The state carries the sword, and is this function of the state not in conflict with Christ's reign of peace?

It cannot be doubted that the eternal

dominion of Christ is a peaceful one. His temporary power, however, he has received with the express purpose of gaining the victory of His Kingdom in continuous struggle. And in that sense we may say that our Lord Jesus Christ also has the power of the sword. But in this connection two remarks have to be made.

In the first place Christ has the power of the sword precisely for the reason that the Kingdom of His grace, the Kingdom of peace, might gain the victory. The power of the sword of the Christ is not foreign to the spirit of His reign of peace, and cannot be construed to contradict it. The Christ who once will judge and crush his enemies is none other than the Christ who offers us His eternal peace. Love does not exist without justice. He will maintain the justice of His love. If one construes this as a contradiction, he shows that he has retained something of the wrong notion of love and justice, namely, the notion that the two contradict each other. Furthermore, he would have to maintain that the kingdom of peace can never enter into this world and must therefore, forever float above it; for if it enters the world it will give rise to struggle there. Christ Himself has said that He did not come to bring peace on this earth, but the sword. So the power of the sword has also been entrusted to him.

We have to add a second remark to this. Christ will not exercise this power of the sword to the utmost until the end of the days. Now, it is true, he executes judgment on this world because of the rejection of his grace, but he still holds this power in abeyance. That means that he holds the power of his grace in abeyance, because when he shall make his grace triumph completely, it will immediately mean the final judgment for all his enemies.

And thus also here we arrive at the conclusion that what typifies the present authority of the state, i.e., the power of the sword, does not form any obstacle to the reduction of this power to the temporary power of Christ.

And with this, I think I have sufficiently developed my second thought. When we state that Christ has dominion over the government we are thinking especially of the temporary power which has been given to him for this period of struggle and for the purpose of the coming of His Kingdom. I do not have to stress any more that this power has been given to him by the Triune God. If at some time I have used the expression that the power of the state is to be reduced to the power of Christ, I did not mean that the source of that power is in Christ. As I have remarked before, this source lies in God, and Christ has received this power from God.

He says that Himself in the well-known saying: "To me has been given all power in heaven and on earth." It is my opinion that in these words he indicated his temporary power, for He adds: "Go ye therefore, and teach all nations." In other words, serve me in the struggle for the complete realization of my Kingdom in the world.

When I say that the state has to serve Christ for the coming of his Kingdom, a further explanation is necessary. I will have to indicate the peculiar character and the limitations of the authority of the state. And thus I have come to the treatment of the third point in which I intend to indicate the consequences of my thesis.

One might be of the opinion that my thesis necessitates upholding Art. 36 of the Belgic Confession in its original edition, in other words, that the government has to avert and to destroy all idolatry and false religion. And one could even entertain the fear that by this train of thought a certain constraint of conscience might be legitimately imposed by the government.

But this danger does not exist at all. The motivation behind the rejection of constraint of conscience and of the original edition of Art. 36 must not be the objectionable thought that the state does not have to serve Christ in His kingly power. This motivation will have to be found in the peculiar character and the limitations of the authority of the state.

The state finds its competency in the jural order as given in the state. The state not only has to maintain the existing legal order, but also is active in creating this legal order (order of justice). Its task is promotion and maintenance of justice. Generally we can say, therefore, that its competency lies in the order of justice.

On the basis of this idea many have the tendency to reject every possibility that the state should be subject to Christ, and should have to serve him in the execution of its task. Some are of the opinion that Christ, when He speaks about His Kingdom, apparently does not attach any value to the order of justice, and does not appreciate its maintenance. Concretely, this would become evident in the well-known part of the sermon on the mount in which He taught that one shall not resist the evil one.

It is strange that about this portion of the sermon on the mount there still exists so much confusion, even while the controversy with the Anabaptists had made it necessary to come to a clear picture of the preceding passage. In that passage Christ just as clearly and absolutely forbids the oath. And concerning the latter we have come to the clear understanding that He forbids every oath which is sworn in one's own interest, but He maintains the swearing of the oath because of the will of God: "To confirm therewith faithfulness and truth." The intent of this part then is completely in agreement with the meaning of the entire sermon on the mount. Christ, in that sermon, takes away our life. We are not our own, but we belong, for the care of our life, to our Father in heaven; and because we are of Him, we are also for Him and unto Him. And thus He develops the law of his kingdom which is not different from the Ten Commandments, but the only correct application of it. In the Kingdom of Jesus Christ we can and must swear the oath because of the will of God. And thus the swearing of the oath is not in conflict with the spirit of the kingdom. And the same goes for the maintenance of justice (law). Christ forbids all resistance to the evil one in which the desire for self-assertion would be the motive force. That one has completely vanquished the desire for self-assertion has become evident in one's willingness to turn the cheek to the one who hit his right cheek. In the Kingdom of Christ the maintaining of oneself for one's own sake belongs to the impossibilities. It is another matter to maintain the right of God because of the will of God. The resisting of the evil one because of God, because the evil one attacks the right of God, cannot be in conflict with the spirit of the kingdom of heaven. That kingdom has come on earth precisely for the reason that the right of God might be restored in this world. I have already argued the power of the sword was therefore entrusted to Christ.

And so we certainly have to say that the state in its power for the promotion and maintenance of justice has to serve Christ. Even a neutral and pagan state serves him even if in spite of itself, that is, despite the intentions of its magistrates. For if the whole legal order as given in the state is not to be understood positivistically, but if it has to be the expression of the divine right, then the existence of the legal order has to be seen as the fruit of the cross of Christ. At this juncture it is impossible to enter upon a defense of this last thesis; this can be done only in the context of an essay on the relation of nature and grace. Nevertheless this last thesis forms the background for the thesis which we propose in this essay. The fact that God did put the dominion over the state also in the hands of Christ stands in direct relation to the fact that the existence of the legal order in the state is the fruit of His cross. By maintaining that legal order, insofar as this order is in conformity with the divine law, the government serves the exalted Christ.

In all this we have not forgotten that the state is able to misuse its power, can worship might for the sake of might, can deem itself free from every higher norm, can choose its own direction, and can commit injustice. Insofar as it is used to maintain something of the divine right on earth it does serve Jesus Christ. Insofar she makes misuses of her power it resists Him. As the prince of the kings of the earth, Christ will also judge the magistrates.

Since it is subject to Christ the government is called to recognize him as the prince of the kings of the earth and to live according to the light of his word, in the fulfillment of its calling.

For the positivation of the right which it has to maintain, and for the determination of the manner in which it has to maintain that right, the state will have to follow the indications in His Word.

This does not serve to deny that the state has to take into account fully the other source of knowledge, viz., the revelation of God in the works of His hands. It will, however, never be able to know this revelation outside of the light of the Word revelation.

The fact that the competency of the state is the legal order gives a strict limitation of her task. It is doubtlessly difficult to determine what does and what does not belong to that legal order. In general we can say that it pertains to the common interest (welfare). It serves for the maintaining of the divine right insofar as the common interest (welfare) does find protection under it. It may not be objected here that God's right and the interest of men as such are contradictory for this objection does not hold if one sees that through Jesus there is again communion between God and this world, and that therefore the maintaining of the right of God implies the well-being of man.

In regard to the character and the limitation of the power of the state we, therefore, have to keep two things in mind. In the first place, it serves for the furtherance of the common welfare, and secondly, it furthers that common welfare only insofar as this functions in the sphere of justice. The terrain of the care of the state is the terrain of the legal relation of the citizens of the state.

Thus the calling of the government cannot be limited to the maintaining of the second table of the law. This arbitrary limitation is impossible. The state has to maintain the entire law of God, but it can do that only insofar as the law of the Lord can find expression in the legal order of the state. The breaking of the third commandment is punished in the Netherlands, and it is punished by the state not as the profanation of the name of God, but, and in my view quite rightly so, because by this profanation the religious feelings of fellow citizens are offended. Thus the state maintains the third commandment insofar as that finds expression in the legal order over which it has to guard.

A moment ago I stated that the state does not only have to maintain the existing legal order, but also participates actively in the formulation of positive justice. It does this in cooperation with parliament. And in this there are two things which it may not forget.

In the first place the state has to reckon with history and tradition. It cannot change the whole law-structure all at once. It may not act in a revolutionary manner, but must reckon with that which has historically grown. All stability would be lacking if it would act differently. Obedience of the state to Jesus Christ cannot therefore imply that it would try to effectuate, if necessary, a complete renewal of the constitution in a short period. Christ is also Lord of history, and it would be a negation of his Kingship if the state would try to interfere in the course of history without due consideration.

In the second place the state has to reckon fully with that which lives among the people. The state consists of government and people. The government does not float above the people, but people and government have been interrelated in the state. If the government would be convinced that the law of the Lord would pose a certain demand for the life of the state, it will be unable to positivize that demand in a law in certain cases, because such a law would not find sufficient response in the conscience of the people. It would be impossible to maintain that law, and the government would thus damage its authority. Thus also the government which is willing to bow before Jesus Christ is thrown into continual compromise. Even in this, one should not see an obstacle to the confession that the government is subject to Jesus Christ. Indeed, the law of the kingdom is always absolute, but Christ does not ask anybody in authority to do anything which falls outside its competency. And it falls outside the competency of the government to push through anything which does not at all find a response in what lives among the people. Thus, in the kingdom of God as it reveals itself on this earth, there is room for a conscious, principial compromise. Principial and compromise are not mutually exclusive. And such a principial compromise means something else than reckoning with the exigencies of practical life.

You will understand that in the development of my third thought I could indicate only a few things. This scantiness was intentional. I am conscious of the fact that as someone who is not a lawyer I have been skating on rather thin ice. You will understand that I preferred to limit myself drastically in this field. The only thing which I felt obligated to show was that, in the maintaining of the kingship of Christ, one hardly can go to the system in our country where no attention is paid to the kingship of Christ in the platforms of the parties. On the contrary, I am of the opinion that for that purpose one has to have a principial Christian political party. In our political struggle we have to be very conscious of the fact that we struggle under the dominion of Him to whom the Father has given all power unto salvation in heaven and upon earth. If the people see to it first that they serve Christ, and march behind the standard of our Saviour, they will be very willing to join the battle also on this terrain.

www.ingramcontent.com/pod-product-compliance
Lightning Source LLC
Chambersburg PA
CBHW052000290426
44110CB00015B/2316